Full STEAM Ahead!

Technology Time

Using Technology Wisely

Cynthia O'Brien

CRABTREE
PUBLISHING COMPANY
WWW.CRABTREEBOOKS.COM

Title-Specific Learning Objectives:
Readers will:

- Identify examples of technology.
- Describe solutions for problems created by technology.
- Identify the reasons the author uses to explain how technology can be helpful and harmful.

High-frequency words (grade one)	Academic vocabulary
can, help, how, make, that, they, this, use, your	battery, energy, environment, pollute, recharge, recycle, technology

Before, During, and After Reading Prompts:

Activate Prior Knowledge and Make Predictions:
Choose three examples of technology in your classroom, such as a computer, a projector, or even a simple pencil sharpener. Present this riddle: What do the three things have in common?

Explain that they are all examples of technology—things people create to solve a problem or make life easier, safer, or more fun. Ask children:

- What kinds of technology do you use?
- What are some ways technology helps us?
- Read the title of this book. What do you think it will be about?

During Reading:
After they read page 7, ask children what the connection is between the two pictures. Help them to see that the top picture presents a problem. The bottom picture shows one solution to the problem. Ask children how the author uses the example to support the idea that technology can be helpful and harmful.

After Reading:
Draw a three-column chart on the board. Ask students to give examples of technology they read or talked about. In the second column, write how each item helps people. In the third column, record any problems caused by that technology.

Author: Cynthia O'Brien

Series Development: Reagan Miller

Editors: Bonnie Dobkin, Janine Deschenes

Proofreader: Melissa Boyce

STEAM Notes for Educators: Bonnie Dobkin

Guided Reading Leveling: Publishing Solutions Group

Cover, Interior Design, and Prepress: Samara Parent

Photo research: Samara Parent

Production coordinator: Katherine Berti

Photographs:
Alamy: Roger Bamber: p. 21
iStock: monkeybusinessimages: p. 10; kirin_photo: p. 11; Stellajune3700: p. 12; FatCamera: p. 19 (bottom)

All other photographs by Shutterstock

Library and Archives Canada Cataloguing in Publication

Title: Using technology wisely / Cynthia O'Brien.
Names: O'Brien, Cynthia (Cynthia J.), author.
Description: Series statement: Full STEAM ahead! | Includes index.
Identifiers: Canadiana (print) 20190231629 |
Canadiana (ebook) 20190231645 |
ISBN 9780778771807 (hardcover) |
ISBN 9780778772651 (softcover) |
ISBN 9781427124586 (HTML)
Subjects: LCSH: Green technology—Juvenile literature. |
LCSH: Technology—Juvenile literature.
Classification: LCC TD145 .O27 2020 | DDC j628—dc23

Library of Congress Cataloging-in-Publication Data

CIP available at the Library of Congress

LCCN: 2019053054

Printed in Canada/112022/CPC20221114

Table of Contents

Crabtree Publishing Company
www.crabtreebooks.com 1-800-387-7650

In Canada: We acknowledge the financial support of the Government of Canada through the Canada Book Fund for our publishing activities.

Published in Canada
Crabtree Publishing
616 Welland Ave.
St. Catharines, Ontario
L2M 5V6

Published in the United States
Crabtree Publishing
PMB 59051
350 Fifth Avenue, 59th Floor
New York, New York 10118

Published in the United Kingdom
Crabtree Publishing
Maritime House
Basin Road North, Hove
BN41 1WR

Published in Australia
Crabtree Publishing
Unit 3 – 5 Currumbin Court
Capalaba
QLD 4157

Technology All Around

Technology is anything people create to make life easier, safer, or more fun. Technology can be tools, machines, or even toys. What technologies do you use?

There are many kinds of technology in your home. Which ones do you see in this picture?

Some toys use lights and sounds and other kinds of technology. They make the toys more fun to use!

The bumps on these soccer shoes make running safer. They stop you from slipping on the grass.

Technology and the Environment

The environment is everything around us. It is important that everyone **protects** it. Using technology can help.

We protect the environment by keeping it clean. Then people, animals, and plants stay healthy too.

We harm the environment when we throw things away. This causes **pollution**.

You can help protect the environment by drinking from **reusable** water bottles. They don't get thrown away.

Cars

Cars are another kind of technology. They make life easier. We use cars to get from place to place. Cars also help us to travel long distances.

Many families use cars to make shopping easier. How else do people use their cars?

Cars are useful, but they can harm the environment.

Cars give off air pollution when they move.

Getting Around

How can we protect the environment from the pollution that comes from cars? Walking or riding a bike helps to keep Earth healthy. The exercise keeps you healthy too!

A bike is a kind of technology that is made for fun. Bikes are also good for the environment.

You can also help the environment by sharing a car ride with friends. Then there is only one car on the road instead of two or three!

Batteries

A battery is a kind of technology that gives **energy**. Batteries make other kinds of technology work. When do you use batteries?

Some toys need batteries to work. This car uses a battery to move.

Flashlights need batteries to light up. They help us to see in the dark.

Many batteries do not last forever. What do you do with old batteries?

Old Batteries

Batteries have **materials** in them that pollute the environment. That means we should never throw batteries into the garbage.

You should **recycle** most batteries. Then the material inside them cannot hurt the environment.

Some batteries can be **recharged**. Cell phones and tablets use batteries like these. They last for a much longer time.

Screen Time

TVs, computers, and video games are kinds of technology we often use for fun. But spending too much time using these technologies can be a problem.

“Screen time” is the time we spend using types of technology with screens. How much screen time do you have each day?

Usually, people sit in one place during screen time. When we sit in front of screens, we are not active. To be active is to move.

Fun with Technology

There are many kinds of fun technology that do not have screens. They help us be active and spend time with family and friends.

It is important that we do not have too much screen time. People need to be active to stay healthy.

Some kinds of technology can help you learn a new hobby.

Technology can help us play games with friends or family!

New Technologies

People are working hard to make new technologies that are less harmful to the environment. They also think of new ways to use old technologies.

Electric cars do not cause air pollution.

People have created toilets, faucets, and shower heads that save water.

People have found ways to reuse parts of old technologies, such as computers. The parts help make new technologies. Reusing items means they will not become trash.

Words to Know

energy [EN-er-jee] noun The power that makes things move, work, and grow

materials [m*uh*-TEER-ee-*uh* lz] noun Things used to make other things, for example, wood to make a table

pollution [poh-LOO-shuhn] noun Something that makes other things, such as air, dirty

protects [pruh-TEKTS] verb To keep from being hurt

recharged [ree-CHARJD] verb To refill with electricity

recycle [ree-SY-cuhl] verb To make something new from something used

reusable [re-YOO-z*uh*-b*uh* l] adjective Able to be used again and again

A noun is a person, place, or thing.

A verb is an action word that tells you what someone or something does.

An adjective is a word that tells you what something is like.

Index

About the Author

Cynthia O'Brien has written many books for young readers. It is fun to help make a technology like a book! Books can be full of stories. They also teach you about the world around you, including other technologies—like robots.

To explore and learn more, enter the code at the Crabtree Plus website below.

www.crabtreeplus.com/fullsteamahead

Your code is:
fsa20

STEAM Notes for Educators

Full STEAM Ahead is a literacy series that helps readers build vocabulary, fluency, and comprehension while learning about big ideas in STEAM subjects. *Using Technology Wisely* helps readers identify the main idea of the book and give examples of how to use technology wisely that support the main idea. The STEAM activity below helps readers extend the ideas in the book to build their skills in technology and engineering.

Technology and Me

Children will be able to:

- Understand what screen time is.
- Discuss problems caused by too much screen time and brainstorm ways to replace it with other activities.
- Create posters that use pictures to promote screen-free activities.

Materials

- Screen-Free Fun Poster
- Crayons or colored pencils
- Screen-Time Challenge Worksheet

Guiding Prompts

After reading *Using Technology Wisely*, ask children:

- What are some examples of technology you read about in this book. How does each one help us? What are some problems caused by each type of technology?

Activity Prompts

Write the following words on the board:

- TV
- Video games
- Computers and tablets

Ask children how much time they think they spend each day using each of these types of technology. Explain that time spent like this is called "screen time" and that some people think too much screen time can be harmful.

Have children do a Think-Pair-Share to discuss what problems they think can be caused by too much screen time. Review pages 16 to 19, guiding students to see that when they are in front of a screen, they are not active, or trying new things, or spending time with friends and family.

Hand out the Screen-Free Fun Poster. Ask each child to make a poster of six screen-free activities that they already do or would like to try. Display the finished posters around the classroom. Have a gallery walk in which children write down their favorite ideas from the posters.

Extensions

- Ask children to take part in a Screen-Time Challenge. Hand out worksheets and tell them that over the next three days, they should look for six chances to trade screen time for another activity. Discuss afterward.

To view and download the worksheets, visit **www.crabtreebooks.com/resources/printables** or **www.crabtreeplus.com/fullsteamahead** and enter the code **fsa20**.